Process Intensification 2017

Topical Conference at the 2017 AIChE Spring Meeting and
13th Global Congress on Process Safety

San Antonio, Texas, USA
26 – 30 March 2017

ISBN: 978-1-5108-4139-0

Printed from e-media with permission by:

Curran Associates, Inc.
57 Morehouse Lane
Red Hook, NY 12571

Some format issues inherent in the e-media version may also appear in this print version.

Printed by Curran Associates, Inc. (2017)

For permission requests, please contact AIChE
at the address below.

AIChE
120 Wall Street, FL 23
New York, NY 10005-4020

Phone: (800) 242-4363
Fax: (203) 775-5177

www.aiche.org

Additional copies of this publication are available from:

Curran Associates, Inc.
57 Morehouse Lane
Red Hook, NY 12571 USA
Phone: 845-758-0400
Fax: 845-758-2633
Email: curran@proceedings.com
Web: www.proceedings.com

TABLE OF CONTENTS

484898 Modularity and the Role of Process Intensification

Abstract:

The recent decades have witnessed significant efforts in both modular production and process intensification. As a common denominator, both process intensification and modular production are predicated on smaller, more efficient plants, where capacity is deployed as needed, and expanded by numbering-up rather than by conventional scale-up approaches. Modularity promotes agility and allows industrial players to enter markets and deliver products faster, with modular plants competing favorably with conventional facilities for certain types of products and market circumstances. In this talk, we discuss modular production in two contexts, centralized and geographically distributed manufacturing, and the impact of the choice of conventional or intensified technology on the economics of modular processes. We provide several examples and identify research needs in the design, control and operation of modular and intensified processes.

482645 Modular Manufacturing: Overcoming Cost and Reliability Barriers for Intensified Chemical Process Equipment

Abstract:

A growing trend in the chemical process industry for reducing capital costs and accelerating chemical plant builds involves building plants on site by assembling together modules, consisting of one or more unit operations, produced at a centralized manufacturing facility. The use of intensified chemical process equipment (ICPE), for reducing the size and weight of unit operations, is consistent with modular manufacturing needs to reduce the costs of producing, shipping and installing modules. However, the cost and reliability of ICPE technologies remains a barrier to the use of these technologies by module manufacturers. Recent manufacturing cost studies show that ICPE heat exchanger technology can be cost competitive with conventional technologies. Further, computational methods are being used to anticipate and overcome reliability challenges. Finally, new manufacturing process technologies will be discussed for further reducing the cost of ICPE.

482649 Process Intensification: AIChE *Rapid* Opportunities in Energy and Chemicals

Abstract:

Challenges in future energy and chemicals include expanded use of natural gas as a feedstock, incorporation of solar and wind into the energy supply, and enhanced concerns over emissions and footprints including land and resource utilization. The newly created ***Rapid*** institute for process intensification (PI) will be a game changer for AIChE in enabling the multi-disciplined collaborations needed to drive sustainable new technologies to commercialization, uniting efforts and capabilities from industry, academia, and national labs. Intensified and/or coupled reactions and separations will drive new paradigms for chemical engineering design to meet cost and efficiency targets. This seminar will provide a look back at both successes and challenges in the implementation of PI to date, and current opportunities for ***Rapid*** in meeting the very significant challenges we face in providing for a sustainable future supply of energy and chemicals.

480908 Process Intensification Toolkit for Mass and Heat Transfer in Rotating Packed Bed

Abstract:

Process intensification (PI) has been widely applied to improve process efficiency on areas such as process capacity and yield, energy consumption, process safety and cost, equipment size, carbon footprint and sustainability, etc. Because of strong shear force that splits liquid into numerous droplets and thin films under a high gravity field, rotating packed beds (RPB) can dramatically increase the mass and heat transfer between gas and liquid. Many research studies [1-3] and industry installations confirm that RPB can reduce equipment volume by 1 to 2 orders of magnitude as compared to conventional packed beds.

With the aim of developing a PI toolkit to support process simulation, research, development, and design of intensified absorption and stripping processes involving RPB, we investigate the effects of gravitational force on gas-liquid interficial areas, mass transfer coefficients, and heat transfer coefficients. The improved understanding on the relationships between gravity and mass and heat transfers allows us to develop a generalied process simulation model for RPB that supports industry's efforts to develop intensified absorption and stripping processes. Examples in the use of the simulation model will be presented.

[1] Yu, C.H., Chen, M.T., Chen, H., Tan, C.S., "Effects of Process Configurations for Combination of Rotating Packed Bed and Packed Bed on CO_2 Capture", Applied Energy, 175, 269-276 (2016).

[2] Yu, C.H., Lin, Y.X., Tan, C.S., "Effects of Inorganic Salts on Absorption of CO_2 and O_2 for Absorbents Containing Diethylenetriamine and Piperazine", Int. J. Greenh. Gas Cont., 30, 118-124 (2014).

[3] Yu, C.H., Cheng, H.H., Tan, C.S., "CO_2 Capture by Alkanolamine Solutions Containing Diethylenetriamine and Piperazine in a Rotating Packed Bed", Int. J. Greenh. Gas Cont., 9, 136-147 (2012).

The Podbielniak® Centrifugal - Applications in Centrifugation

Timothy J. Pomaville
B&P Littleford, LLC
Saginaw, MI, USA

Abstract

Through continuing research and laboratory testing, the Podbielniak® Centrifugal, widely known as the Pod®, has proven itself time and again over broadly differing process applications. Some of the most renowned applications for the Pod, including the streamlining of the penicillin production process, the extraction and concentration of metal ions, as well as many applications in the chemical, petroleum and other industries are described herein along with a case in point to accompany each application. Though there is no new research presented here, recent testing has brought about a shift toward new research of solid phase separation applications to improve upon existing applications. After nearly a century, the Pod continues to have unrealized potential.

A brief discussion of operational theory is also provided in simple form to acquaint the reader with the operational value of the Pod Centrifuge in industry.

Keywords: Pharmaceutical, Chemical, Coke By-products, Hydrometallurgy, Petroleum, Petrochemical, Washing, Separation, Minerals, Mining, Fermented Broths, Solids Handling

Introduction

During the early 1950s, the Pod contactor saw its first, and arguably, one of its most prolific applications take hold within the pharmaceutical industry for the concentration and refining of penicillin. Since this time, the majority of the world's penicillin has been extracted from fermentation broths using the Pod contactor. Over the years, the Pod's numerous applications emerged in petroleum, soap, chemical and metallurgy industries as well. Within the petroleum industry, Pods have been used to replace gravity towers with comparable capacities, greater efficiencies and greater operational flexibility. The description that follows is a simplified explanation intended to lift the shroud of mystery and ignite the reader's imagination with past, present and future applications.

Figure 1: Pod Contactor

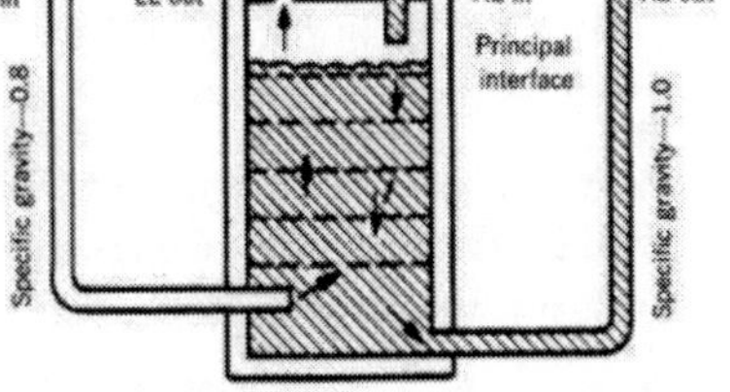

Figure 2: Tower Analogy of a Centrifugal Extractor

Principal of Operation

From an operational point of view, centrifugal force induces the counter-current flow of two liquid phases. The key to controlling an extraction or separation is through control of the principal interface. This is most readily visualized using tower analogy, a comparison to perforated plate, bubble tray or down-comer gravity column to better understand the most important features of the Pod's operation. If the back pressure is increased on the light liquid out stream, the interface will move away from the exit conduit, allowing for more hold-up of the lighter phase within the Pod. In this way, the Pod works the same way a gravity column does, but the centrifugal force generated can be 100s of times that of a gravity column. Like any other liquid/liquid extraction unit operation, the Pod is then characterized by the number of theoretical stages calculated for a given application.

What's Physically Happening Inside?

A – Heavy liquid out (HLO) - leaves the rotor after being centrifuged to the outer rim.

B – Light liquid out (LLO) – leaves via the shaft after being displaced inward.

C – Light liquid in (LLI) enters the rotor slightly inboard of the rim and the HLO.

D – Heavy liquid in (HLI) enters an outboard from the shaft. It is then centrifuged outboard to the rim, displacing the light liquid, inducing the counter-current flows.

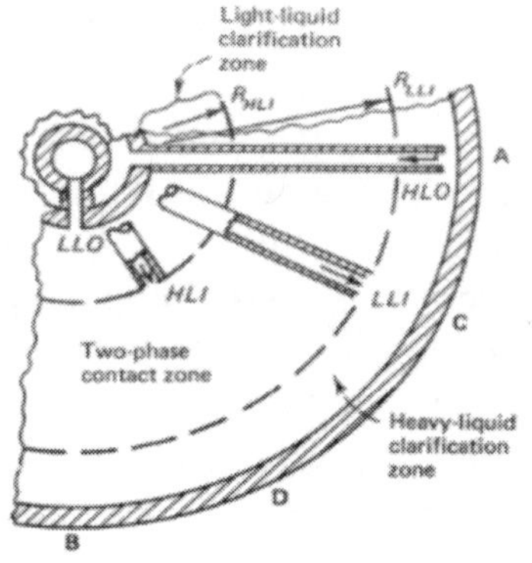

Figure 3: Cross-section of a Pod rotor

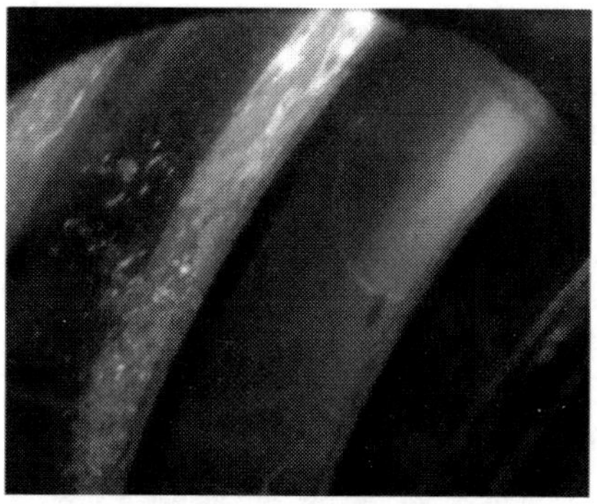
Figure 4: Window rotor Pod in operation

The internal fluid dispersion shown in Fig. 3 illustrates the above concept. In general, to observe the intended effects of a Pod, there must be a differential in the specific gravity of at least 0.01. This difference in specific gravity allows for the manipulation of the interface through adjustment of pressure differential. A continuous phase can be chosen for the whole rotor if the process dictates the need. With the interface near the shaft, the dispersed phase will be the light liquid or light key. Conversely, if the interface is near the rim, the dispersed phase will be the heavy liquid or heavy key. Selection of perforated element bands, along with careful positioning of the fluid entry points, affords flexibility from one application to another.

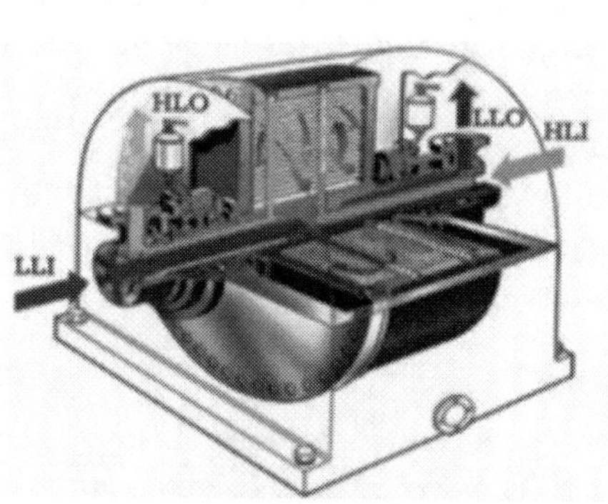

Figure 5: Flow through a Pod

For simplicity and brevity, imagine the washing operation of kerosene oil with water. This is a classic example which is used repeatedly for demonstration purposes. Water, the heavier liquid is introduced to the Pod to lubricate and keep cool the hermetic mechanical seals. The rotor is started, and kerosene oil is introduced to the system. A back-pressure is then applied to the light liquid out (LLO) stream. Heavy liquid out (HLO) increases momentarily as a portion of the volume of water within the rotor is displaced. At this time, the principal interface begins to move outboard, away from the shaft. After a short time, the system stabilizes and reaches a steady state. Light liquid enters near the rim, leading rotation and passing through the heavy liquid of the heavy liquid clarification zone by way of displacement due to the specific gravity difference in the form of droplets. Small droplets (higher surface area) expedite the extraction of impurities from the kerosene oil into the water phase. Simultaneously, heavy liquid enters the system just outboard of the shaft, lagging rotation and light liquid clarifying zone. Droplet formation is facilitated by the bands, exposing surface area of the heavy phase, allowing impurities to pass into the water or solvent phase. The water, containing impurities, is sent outboard to exit at the periphery of the rotor (rim).

Pharmaceutical Pod Applications

The Pod's earliest successes were seen in the pharmaceutical industry, as evidenced by high volume production of antibiotics such as penicillin and other fermentation products. In fact, problems that were particular to large-scale production of penicillin heavily influenced the development of the Pod. As the pharmaceutical industry grew, the Pod evolved to accommodate growing fermentation capacities until a full line of standardized Pods were developed with features that emphasize easy maintenance and cleaning. Certainly, these developments have been utilized in other industrial markets, but the Pod has matured through the pharmaceutical industry and maintains a significant role in its extraction processes.

Fundamental to the maximum recovery of unstable products from low pH, non-refrigerated broths, as in the case of penicillin production is a swift contacting time from the fermenter to the finished product. The Pod lent itself well to solving these problems with commercial scale pharmaceutical production that were previously not possible. Since the potency of the acidulated broths is time dependent, the fraction-of-a-minute holdup time permitted maximum recovery of product potency despite its relatively short half-life. Additionally, many broths that show extreme tendencies to emulsify due to the many active agents added to their respective fermentation processes can be recovered in a Pod.

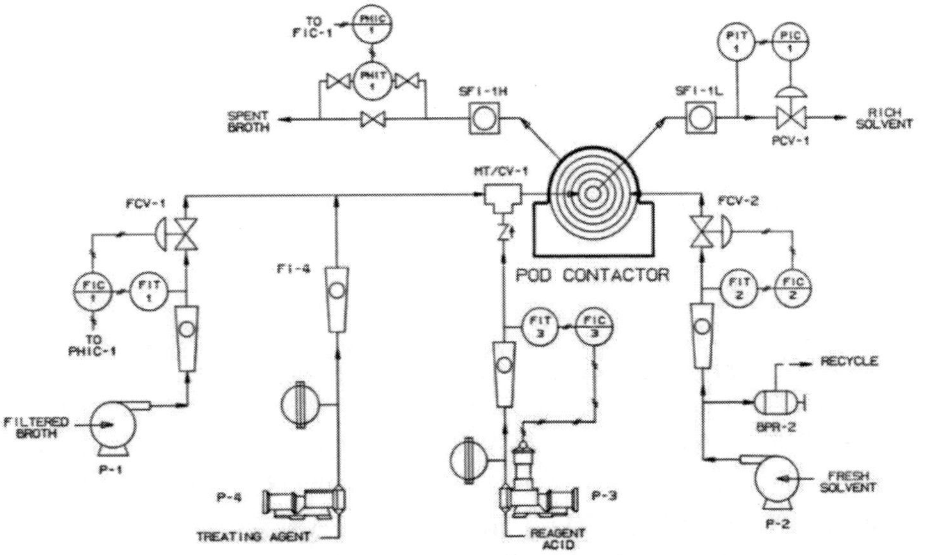

Figure 6: Primary Antibiotic Extraction

Although it is of course process dependent, Pods have seen upwards of five theoretical stages within one unit, contributing to 98% or better extraction efficiency in a single step. Selective solvent to broth ratios were minimized, permitting maximum concentration of products and effectively reducing solvent inventories. At the same time, high G-forces imparted by the Pod permitted maximum clarity to the product streams.

These advantages specific to the penicillin process became widely used in industry for the production of antibiotics such as Bacitracin, Cloromycetin, Kanamycin, Tetracyclin, Estrogens, Steroids, Cortisone, Ephedrine, Opiates, a variety of vitamins and many others. Large scale Pods were subsequently developed and standardized for the production of pharmaceuticals to process greater volumes of fermented broth. The stream arrangement shown Fig. 4 is the most common in pharmaceutical applications. Many pharmaceutical producers using Pods practice counter-current extraction on one or two Pods working in series with only slight variations that depend on broth quality at a given time. In some cases, producers have used special techniques tailored in-house to meet their own, custom applications.

Petrochemical Pod Applications

Waste streams from refineries containing fractions of naphthenic acids can be removed by treating with caustic. The sodium naphthenates formed in the process can be easily disposed or reused to regenerate naphthenic acids for further use. A simplified flow diagram for the recovery of naphthenic acids from refinery waste streams is illustrated below in Fig. 6. Spent caustic is mixed in-line with a strong inorganic acid and resulting in a weaker organic acid. The emulsified mixture enters the Pod to be washed with water for the removal of impurities. Recoveries of naphthenic acid in excess of 99% have been observed. The salt water, with acceptable impurities, can be sent to the sewer. The pH of the salt water stream can be continuously analyzed to easily automate changes in the feed ratios.

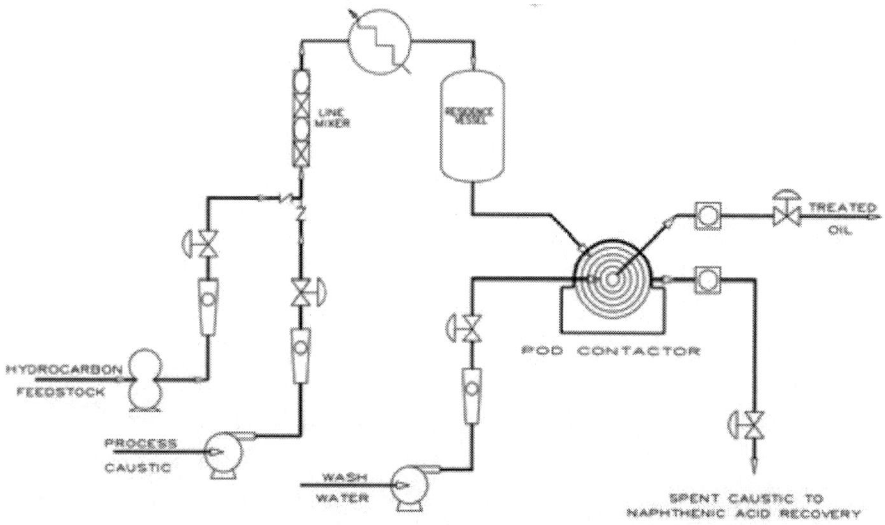

Figure 7: Caustic Treat/ Water Wash to Remove Nap Acids

Solvent extraction from lubricating oils was proven decades ago in the petroleum industry. A single pass through the Pod using a selective solvent produces a raffinate of outstanding quality. Undesirable components are removed and rejected to the extract stream where they can be stripped and reused for subsequent extraction. As compared to a commercial perforated tray column, the Pod requires only 1/45 of the holdup with reduced off spec production when changing from one grade of lube oil to another.

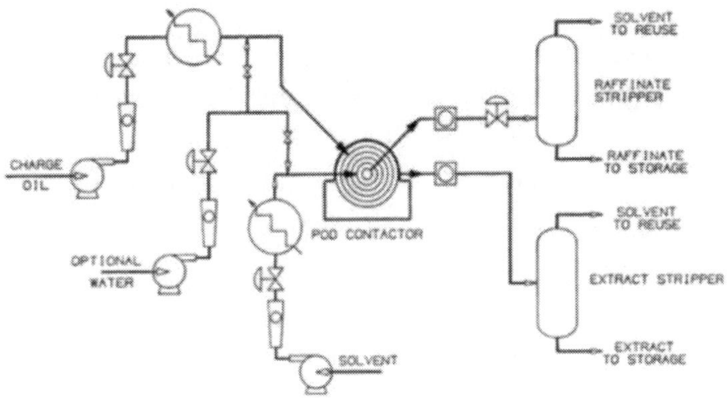

Figure 8: Simplified Solvent Lube Oil Extraction

Within the petrochemical industry, Pods can also be used to extract aromatics and coke-forming compounds in connection with catalytic cracking operations. Cycle stock would be used to raise crude charge stock capacity, catalytic cracker charge capacity and percent yield of gasoline. Use of a Pod as an extractor reduces coke burning requirements, increasing gasoline yield and reduces catalytic cracker investments.

The advantages are numerous and similar to those previously mentioned for other applications. Controlled reactions, solvent extractions and separations are all accomplished with a Pod in the continuous sulfonation application, for example. Shorter processing times for the separation of sludge from an oil/sulfonate mixture are accomplished in minutes as opposed to hours. Light colored sulfonates are produced due to the elimination or overreacting or "burning." The Pod makes the best use of oleum with no side reactions, such as unstable or off color products, to name a few benefits.

Hydrometallurgical Pod Applications

Pods have been used extensively for the extraction, concentration and purification of molybdenum, zirconium, vanadium and titanium metal complexes, as well as rare earths such as thorium and cesium through the use of selective solvents. Pods are especially useful in recovery of regenerative complex liquid ion exchange mediums that are dissolved in solvents immiscible with aqueous solutions. In fact, the bulk of practical solvent extractions involve complex formations as a way of influencing the distribution in the system.

In hydrometallurgical applications, an aqueous solution containing the desired metal is contacted with a suitable solvent. The metal, in some form, will distribute between the aqueous and organic phases. To make the metal favor the organic phase, an extractant or reagent is often used which can react with the metal to form complexes. These metal complexes are the predominant species that are extracted into the organic phase. For hydrated metal ions existing in aqueous solution which cannot be extracted to a significant extent, organic solvents are reacted to replace the water of hydration. This is accomplished by combining the metal ions with ions of an opposite charge to form a neutral metal species. The larger and more hydrophobic the resulting species, the better the extraction will be.

An especially useful application for Pods is in the extraction of rare earths from sulfate, nitrate or chloride solutions. Though some of these processes have certainly evolved, a 15 second pass through a Pod has demonstrated that at 500 GPM of pregnant sulfuric acid leach liquor using 80 GPM organic solvent can yield up to 0.8 gpl of rare earth metals (loading step). A single pass through a smaller Pod (stripping step) transfers rare earth salts into an ammonium nitrate solution for the recovery of these various rare earth metals by precipitation into an appropriate solution. The organic solvent is then recovered via scrubbing and recycled back to the first step to complete the loop.

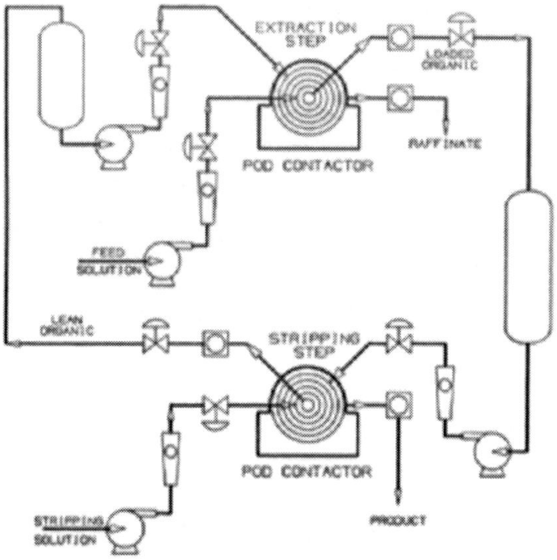

Figure 9: Simplified Hydrometallurgical Liquid Ion Exchange

A conventional mixer-settler plant would need to use 60,000 gallons of solvent (5-10% solution of Amine complexing agents), while a Pod would require only 3,000 gallons to accomplish the same task. This amounts to a 20 fold decrease in solvent inventory. With its multiple theoretical stages, solvent extraction in one Pod has been seen to leave only one ppm of rare earth metal in the waste liquor. Automated instrumentation assures close control of the process with minimal operator interaction. The control system is actuated by changes in the flow of pregnant liquor and ratio controllers adjust other flow rates accordingly.

Pods, here again, have the advantage of up to 5 theoretical stages for the extraction of metals and the use of centrifugal force to disengage the phases. Because of the short hold-up time within a Pod, they can be used to separate two or more metals contained in a given solution when the reaction rate for one metal is considerably shorter than another as opposed to columns or mixer settlers. Varying the RPMs or throughput can permit improved selectivity in the separation of a desired metal

Upcoming Research

A significant interest in many Pod applications arises when microscopic solids are introduced to the system. There are many existing Pod applications that would see improvement through shedding of solids from the internals, especially when shutting down the process is a drawback to the user. A solution will require extensive research that may need to be evaluated for several months to arrive at any definitive conclusions. Some suggestions have been proposed to keep solids from depositing, but much of the work has yet to be attempted. Promising results in testing have recently provoked a new interest in this behavior as it relates to solids in the Pod. Using a system of two immiscible liquids, one of them containing particularly challenging solids, research to smooth the progress of existing and promising new applications are being thoroughly investigated.

Conclusion

The Pod centrifugal contactor has extraordinary processing capabilities and a well-built mechanical design. They have been used broadly and successfully in many applications and continue to have untapped potential in many other untested applications. For these untested applications, it may be necessary to conduct laboratory testing for specific materials or process requirements to properly evaluate operating conditions and performance. While it is possible to test at off-site facilities, feasibility and pilot testing in the Saginaw, Michigan facility is available for a wide variety of different process configurations along with all of the auxiliary equipment required of commercially standardized models.

Summary of Benefits

The performance of the Pod is unique in its applications. It has become generally accepted that a single unit provides up to five theoretical stages of counter-current contact. One Pod has sufficiently high efficiency to achieve phase separation and counter-current washing of effluent product streams. The largest Pod available today can achieve up to 600 GPM of total combined flow. They are compact, easy to operate and readily cleaned and transported. Materials of construction are stainless steel and other corrosion resistant alloys. Liquid holdup within a Pod ranges from 1-260 gallons depending on the model. Liquids can be passed through a Pod in a fraction of a minute, and if longer residence times are required, liquid flows can be easily adjusted or recycled. Steady state conditions are reached quickly upon start-up and small continuous runs can be completed quickly and effectively with negligible transitional fractions. Pods are inherently pressure tight, closed systems, handling up to 250 psig and up to 600 psig by special design. In addition to being pressure tight, they are suitable for operation from below 0°F - 250°F and 450°F by special design. They are low maintenance with exceptionally long life expectancies.

References

[1] Davies, G.R., Kaiser, H.R., Todd, D.B., "Controlled Centrifugal Separation of Liquids": Baker Perkins Inc., Saginaw, Michigan

[2] Todd, David B., Davies, Gordon R., "Performance of Centrifugal Extractors": International Solvent Extraction Conference, Lyon, Vol.3, 2380 – 2398, September 1974

[3] Todd, David B., Davies, Gordon R., "Centrifugal Pharmaceutical Extractions": Filtration & Separation, November/December 1973

[4] Anderson, S.O., Ashbrook, A., etal, "Solvent Extraction in Process Metallurgy": University of Waterloo and Mines Branch, September 1971

[5] Todd, David B., Podbielniak, W.J., "Advances in Centrifugal Extraction": Chemical Engineering Progress, May 1965

2017 AIChE Spring Meeting Presentation
Paper 99c, Abstract #481626

CRITICAL PARAMETERS FOR DESIGN, TROUBLESHOOTING AND OPTIMIZATION OF AN ETHERIFICATION PROCESS

Minghua Ye, Sergio Mohedas, Brian Peng, and Chih Ning Kuan

March 28, 2017

CHEMISTRY THAT MATTERS™

REACTION: REVERSIBLE REACTION

Low temperature synthesis: **Tertiary Olefins + Alcohols → Ethers**

High temperature back cracking: **Ethers → Tertiary Olefins + Alcohol**

Application:

- Convert low value methanol and ethanol to high value fuel
- Improve octane number for gasoline fuels
- Produce high purity isobutene (incorporate the MTBE back-cracking process)

Typical ethers:

- MTBE (Methyl Tertiary Butyl Ether)
- TAME (Tertiary Amyl Methyl Ether)
- ETBE (Ethyl Tertiary-Butyl Ether)

No. 1

CONCEPTUAL ETHERIFICATION PROCESS

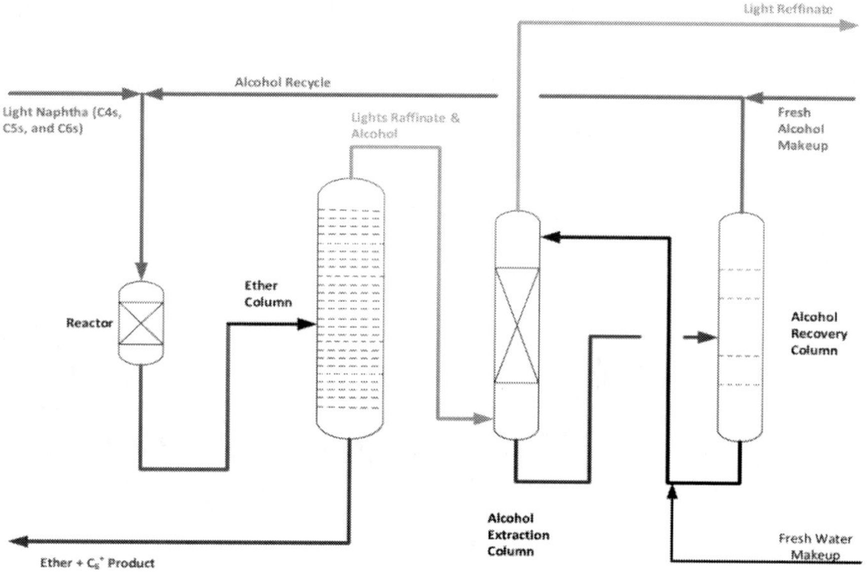

No. 2

ADDING FEED PRETREATMENT UNIT TO REMOVE CONTAMINANT (PROTECT CATALYST)

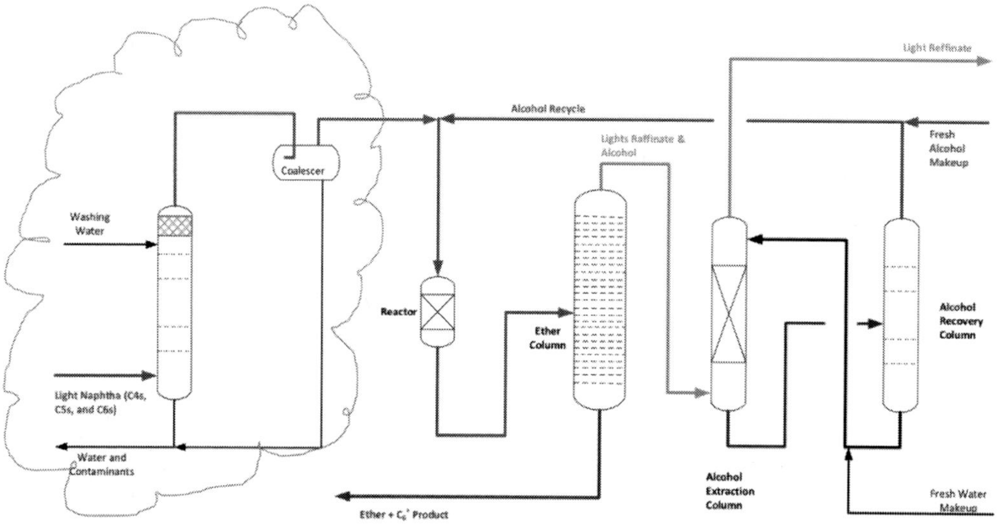

No. 3

INCORPORATION OF REACTIVE DISTILLATION TECHNIQUE TO IMPROVE OVERALL CONVERSION

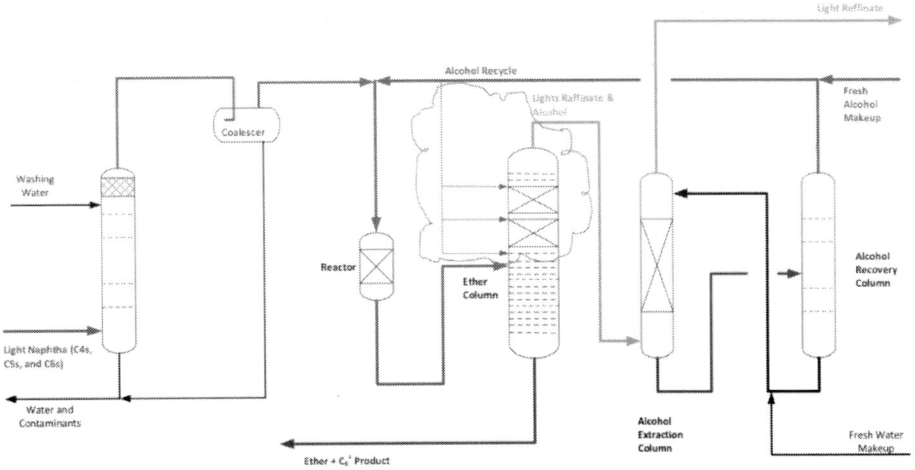

CRITICAL PARAMETERS FOR PROCESS DESIGN, TROUBLESHOOTING AND OPTIMIZATION

- Molar ratio of alcohol to reactive olefin
- Etherification reactor operating conditions
- Azeotrope / reactive distillation
- Effective recovery of alcohol

KEY POINTS

- Feed quality
- Reactor conversion / selectivity
- Reactive distillation column conversion / selectivity
- Overall conversion / selectivity
- Alcohol consumption
- Utility and energy consumption

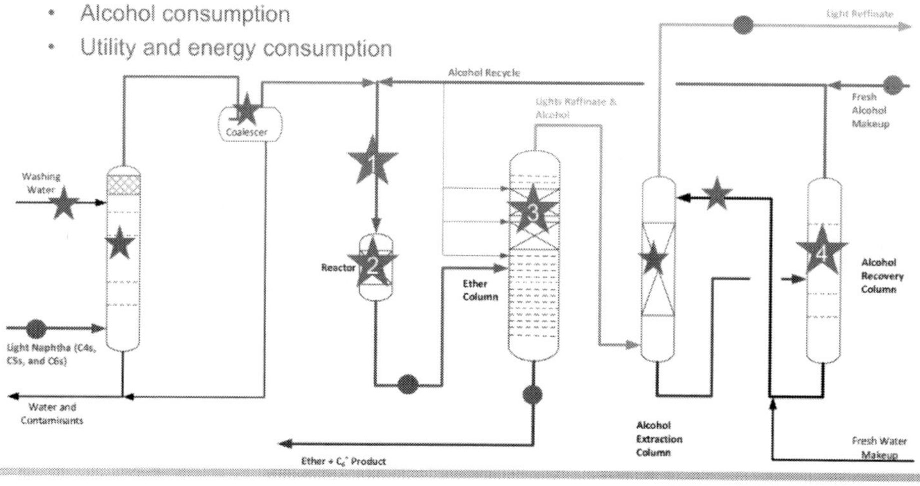

No. 6

PROTECT CATALYST: FEED PRETREATMENT

Catalyst Deactivation

Permanent:

- Nitrogen compounds (such as acetonitrile)
- Metal Ions (Ca^{++}, Mg^{++}, Fe^{+++}, NH_4^{+})
- Catalyst fouling by polymer forming agents (such as butadiene, pentadienes)

Temporary: High water content in feed (resulting in the loss of sulfonic acid groups)
Water may react with tertiary olefin: $iC_4H_8 + H_2O \rightarrow$ TBA (Tertiary Butyl Alcohol)

Impact:

- Reduce catalyst activity
- Shorten the catalyst life
- Increase reactive distillation column loading and operation

Feedstocks:

- The reactive olefins vary from different operating mode of upstream units.
- The contents of impurities must meet specification set by catalyst vendor.

No. 7

MOLAR RATIO OF ALCOHOL TO REACTIVE OLEFIN

$$R = \frac{Alcohol\ Molar\ Flow\ Rate}{Total\ Molar\ Flow\ Rate\ of\ Reactive\ Olefins}$$

Reactive Olefins are the tertiary olefins.

High alcohol ratio can result in alcohol dehydration, for example:

$CH_3OH + CH_3OH = CH_3\text{-}O\text{-}CH_3 + H_2O$

DME formation must be limited especially because:

- it is a very volatile compound and will impact the RVP, an important index of gasoline
- It is a poison in alkylation process

Low alcohol ratio can result in:

- Low reactive olefin conversion
- Dimerization of reactive olefins: 2 i-C_4H_8 → dimer
- Increment of reactive distillation column loading

MAIN REACTIONS IN THE ETHERIFICATION PROCESS

Methanol + tertiary olefins to form ethers

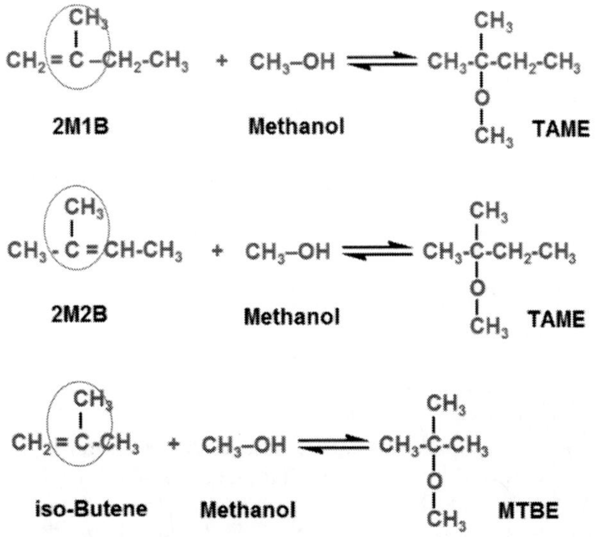

REACTIVE OLEFINS – TERTIARY OLEFINS

C4 Olefin	isobutene		+ CH_3OH ===	Methyl tertiary butyl ether (MTBE)
C5 Olefin	2-methyl-2-butane	2-methyl-1-butane	+ CH_3OH ===	2-methyl-2-methoxy butane (TAME)
C6 Olefin	2-methyl-1-pentene	2-methyl-2-pentene	+ CH_3OH ===	2-methyl-2-methoxy pentane
	cis-3-methyl-2-pentene	trans-2-methyl-2-pentene	+ CH_3OH ===	3-methyl-3-methoxy pentane
	2-ethyl-1-butene			
	2,3-dimethyl-1-butene	2,3-dimethyl-2-butene	+ CH_3OH ===	2,3-dimethyl-2-methoxybutane
	1-methyl-cyclopentene		+ CH_3OH ===	1-methyl-1-methoxy cyclopentane
C7 Olefin	2-methyl-1-hexene	2-methyl-2-hexene	+ CH_3OH ===	2-methyl-2-methoxy hexane
	cis-3-methyl-2-hexene	trans-3-methyl-2-hexene	+ CH_3OH ===	3-methyl-3-methoxy hexane
	cis-3-methyl-3-hexene	trans-3-methyl-3-hexene		
	2-ethyl-2-pentene			
	2-methyl-1-pentene	2-methyl-2-pentane	+ CH_3OH ===	2,3-dimethyl-2-methoxy pentane
	cis-3.4-dimethyl-2-pentene	trans-23,4-dimethyl-2-pentene	+ CH_3OH ===	2,3-dimethyl-3-methoxy pentane
	2-ethyl-3-methyl-1-butene			
	2,4-dimethyl-1-pentene	2,4-dimethyl-2-pentene	+ CH_3OH ===	2,4-dimethyl-2-methoxy pentane
	3-ethyl-2-pentene		+ CH_3OH ===	3-ethyl-3-methoxy pentane
	2,3,3-trimethyl-1-butene		+ CH_3OH ===	2,3,3-trimethyl-2-methoxy butane
	1-ethyl cyclopentene		+ CH_3OH ===	1-ethyl-1-methyoxy cyclopentane
	1,2-dimethyl cyclopentene	1,5-dimethyl cyclopentene	+ CH_3OH ===	1,2-dimethyl-1-methoxy cyclopentane

Ref: Halim Hamid Mohammad Ashraf Ali, Handbook of MTBE and Other Gasoline Oxygenates, Marcel Dekker, Inc, pp 269

SIDE REACTIONS

(1) Methanol Dehydration
$2\ CH_3OH \rightarrow DME + H_2O$
Kinetically favored by high temperature, high concentration of methanol
DME is a very volatile compound that leaves the process with C_4 stream
DME is a poison in alkylation

(2) Dimerization of isobutene to give C_8 hydrocarbons
$2\ iC_4H_8 \rightarrow$ Dimer (C_8H_{16})
Favored by low methanol concentration
Leave the process unit with MTBE, decreasing MTBE purity.

(3) Isobutylene Hydration for tert-Butyl Alcohol
$i\text{-}C_4H_8 + H_2O \rightarrow TBA$
Reaction is promoted by the quantity of water present
Affecting MTBE purity, catalyst activity, and overall conversion

(4) Normal butylene etherification:
$N\text{-}C_4H_8 + CH_3OH \leftrightarrow$ 2-methoxy butane
Kinetically limited, very low conversion, favored at high temperature
2-methoxy butane have much lower octane number than MTBE
Cannot be separated from MTBE

ETHERIFICATION REACTOR OPERATING TEMPERATURE

Exothermic and equilibrium limited, liquid phase reaction
Fixed bed reactor is commonly used in industry
Lower T: Require more catalyst, longer catalyst life, lower DME and Dimer formation
High T: Require less catalyst with shorter life, higher DME and Dimer formation

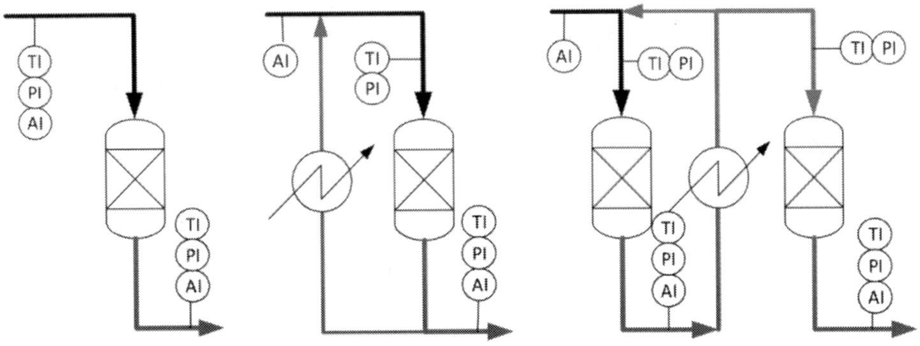

ETHERIFICATION REACTOR OPERATING PRESSURE

The operating pressure depends on the reactor type, operating temperature, and reactor outlet pressure required to deliver the product to downstream units.

- Liquid Phase Reaction (no vaporization in the reactor):
  P_{out} > bubble point pressure at reactor exit temperature.
  P_{out} > down stream separation unit set pressure + pipeline pressure loss
  $P_{in} = \Delta P_{bed} + P_{out}$

- Vaporization Fixed Bed Reactor (reaction heat is removed by partially vaporizing the liquid mixture)
  P_{out} < bubble point pressure at reactor exit temperature
  P_{out} > down stream separation unit set pressure + pipeline pressure loss
  $P_{in} = \Delta P_{bed} + P_{out}$

REACTIVE DISTILLATION (ACHIEVING HIGH ETHERIFICATION CONVERSION)

Thermodynamics:

- highly non-ideal VLE and LLE for HC, alcohol, and ether mixture system
- Low temperature azeotropes of alcohol-HC form in the reactive distillation columns.

Operating parameters

- Alcohol injection – Promote further etherification in the catalyst bed zone
- Overhead HC + alcohol, azeotrope with lower temperature boiling point
- Column operating T, P, and reflux rate

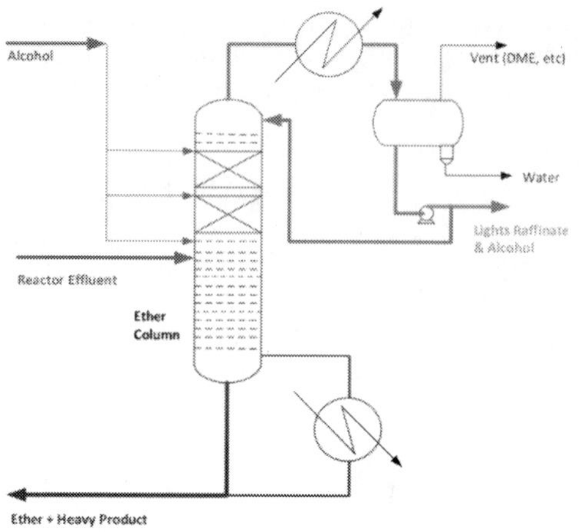

ISOBUTENE + METHANOL + MTBE BINARY VLE DIAGRAMS

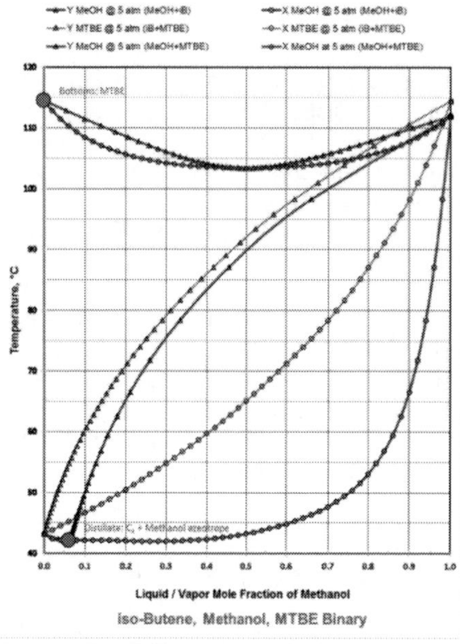

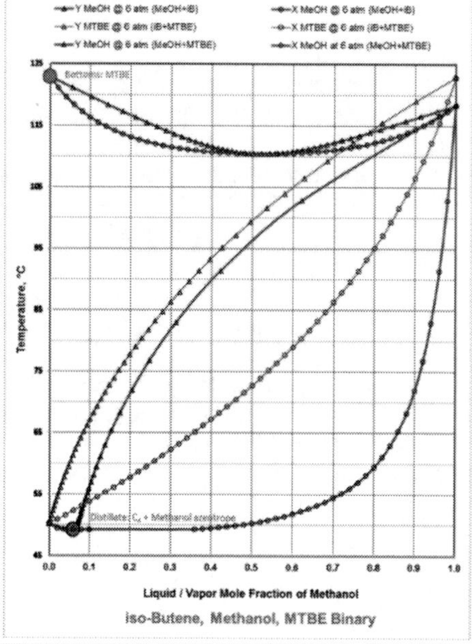

2-METHYL PENTENE-1 + METHANOL + MTPE BINARY VLE DIAGRAMS

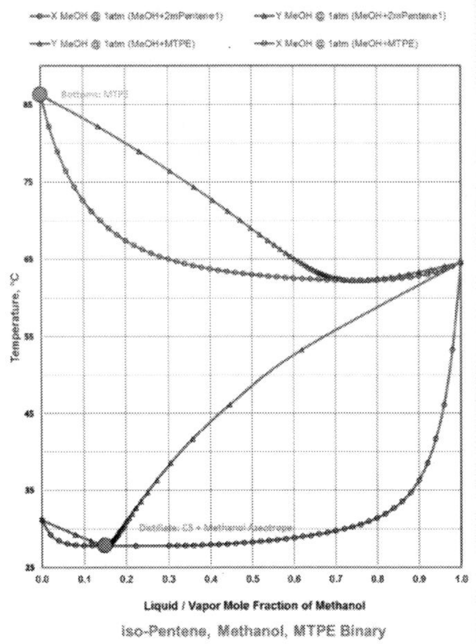

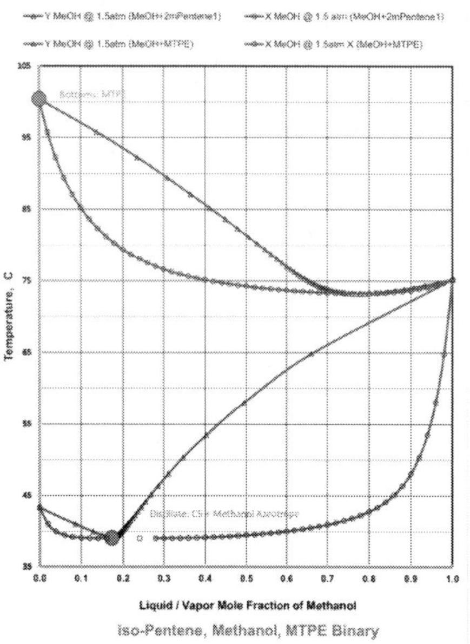

ALCOHOL EXTRACTION

Excess alcohol recovery:
- Liquid extraction of alcohol with water
- Provide sufficient theoretical stages (number of trays or the height of structured packing) to minimize washing water amount
- Use optimal amount of washing water

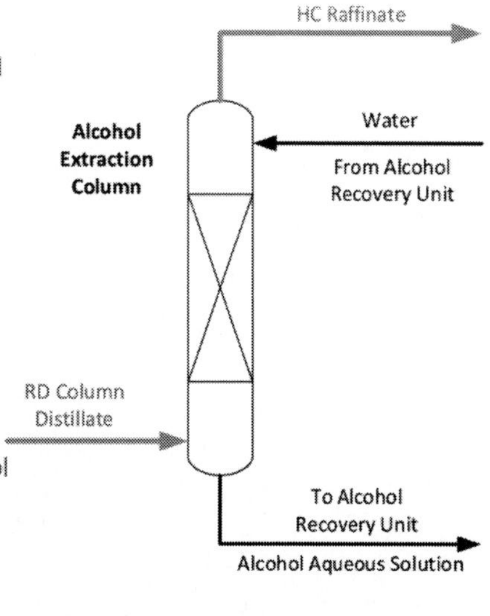

Excess water could results in:
- Larger extraction column size
- More energy to separate alcohol from the aqueous solution
- Larger size for the equipment in the alcohol separation unit (distillation column, condenser, reboiler, drums, etc.)

HIGH PURITY ALCOHOL RECOVERY

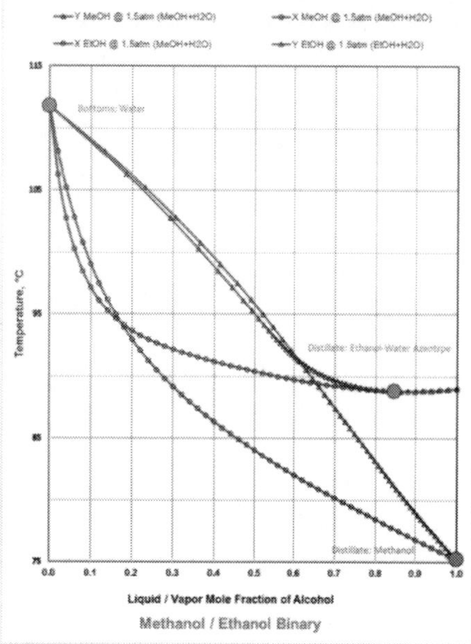

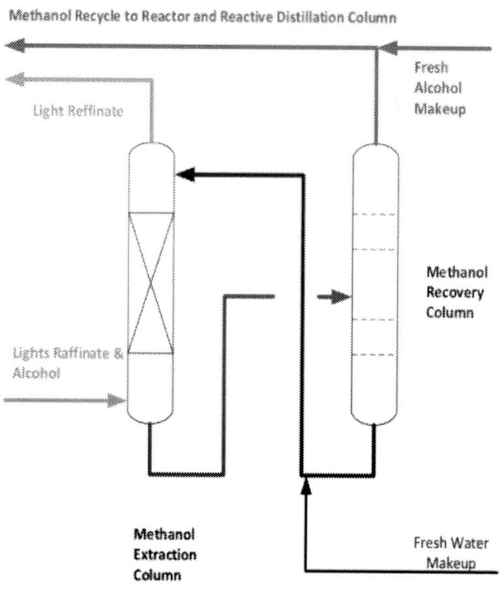

REQUIREMENT FOR ALCOHOL PURITY

- The etherification reactor and reactive distillation bed require water content in feed to be less than 1000 ppm. Both the recovered and fresh alcohol need to be monitored to ensure the recycle alcohol quality.
- For methanol recovery unit, it is achievable by distillation to obtain 99.9 wt% methanol.
- For ethanol, additional extractive distillation technique may be applied to obtain high purity ethanol.
- Alcohol content in the bottoms (water) to be well controlled. High alcohol content in water result in
  - ➔ Alcohol loss to raffinate (extraction column overhead)
  - ➔ Alcohol loss to purge stream

SUMMARY

The critical parameters for etherification process:

- Feed pretreatment to remove catalyst poisons from feed
- Set optimum molar ratio of alcohol to reactive tertiary olefins prior to reactor
- Optimal etherification reactor operating conditions to compromise reasonable reaction kinetic rate and equilibrium conversion
- Reactive distillation for etherification involves formation of azeotropes of HC-alcohol, optimum alcohol injection amount ensures the etherification reaction takes place at low temperature (bubble point of HC-alcohol azeotrope). Excess methanol will result in high alcohol content in the ether product, and higher energy consumption for alcohol recovery
- Effective recovery of alcohol (extraction and distillation) is important to achieve high energy efficiency, recycle alcohol water content, overall reactive olefin conversion, and the overall unit performance

No. 20

DISCLAIMER

No. 21

2017 AIChE®

SPRING MEETING&

13TH

GLOBAL CONGRESS

ON PROCESS SAFETY

March 26–30, 2017
Henry B. Gonzalez
Convention Center
San Antonio, TX

17AIChE
Spring Meeting, San Antonio, TX

13TH GLOBAL CONGRESS ON PROCESS SAFETY

Integrated LPG Upgrading Process Using Divided Wall Columns

Scott Schroeder, Advisian
scott.schroeder@advisian.com

Armen Abazajian, Advisian
armen.abazajian@advisian.com

www.advisian.com

Advisian
WorleyParsons Group

Process Intensification

A philosophy of design that seeks to bring dramatic improvements in the process efficiency, equipment size, or waste reduction through the use of process-intensifying equipment or process-intensifying methods.

Process-intensifying equipment – novel reactors such as micro-channel reactors or rotating packed-bed reactors; novel heat exchangers such as micro-channel heat exchangers

Process-intensifying methods – such as hybrid separations, integration of reaction and separation, integration of heat exchange and separation

Process Intensification

The field of Process Intensification has developed a large tool set of equipment and methods

This tool set includes mature technologies that pre-date the term "Process Intensification" as well as new cutting edge technologies.

Process Intensification - DWC

One traditional method of process intensification is the integration of distillation sequences.

Distillation is the one of the most common separation operations used in the process industry. It accounts of a large portion of the energy usage. Reduction of energy usage has been a key driver in developing integrated distillation sequences.

Divided Wall Columns (DWC) take the concept integrating a distillation sequence to the equipment level.

Conventional Column Sequences

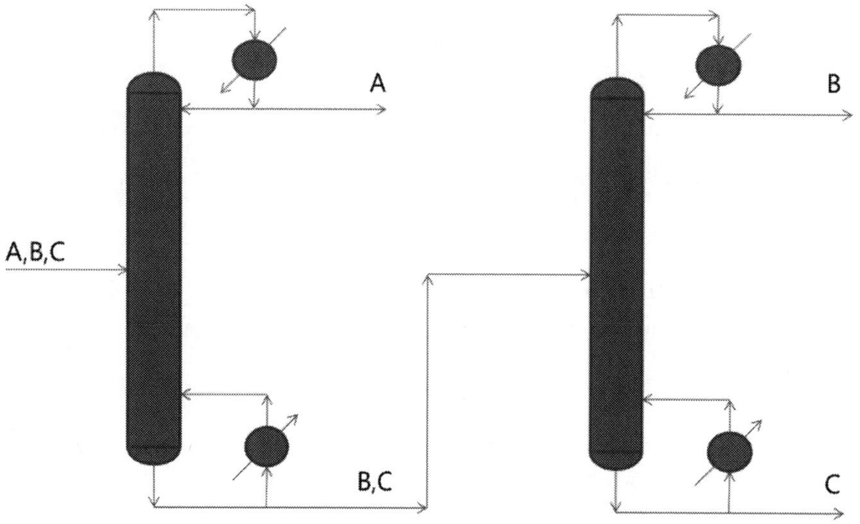

Conventional Column Sequences

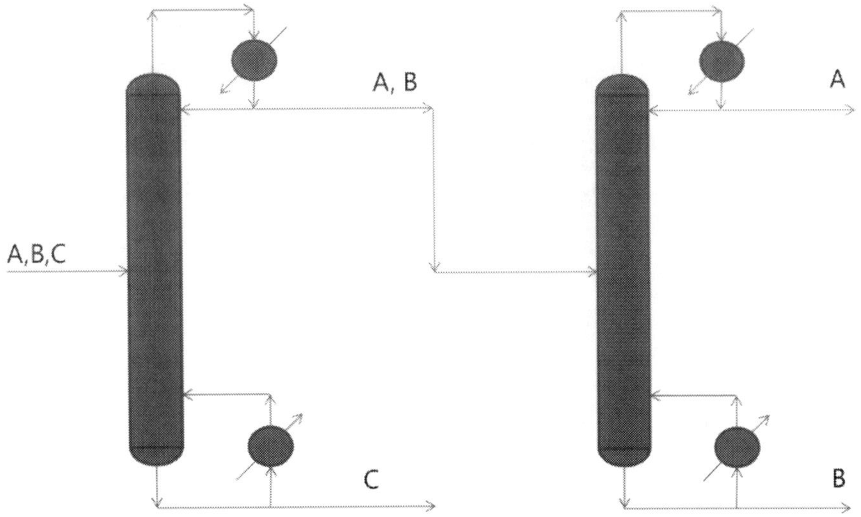

2017 AIChE® SPRING MEETING & 13TH GLOBAL CONGRESS ON PROCESS SAFETY

Advisian / 7

Petlyuk Column

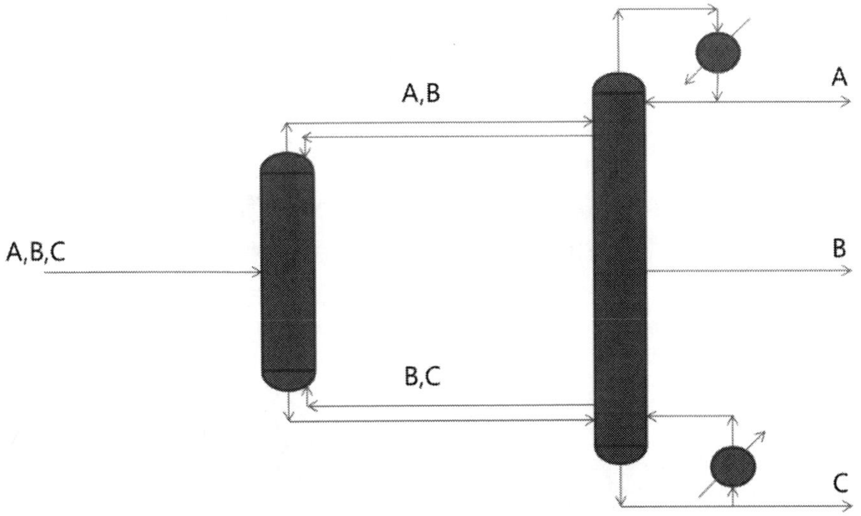

2017 AIChE® SPRING MEETING & 13TH GLOBAL CONGRESS ON PROCESS SAFETY

Advisian / 8

Divided Wall Column (DWC)

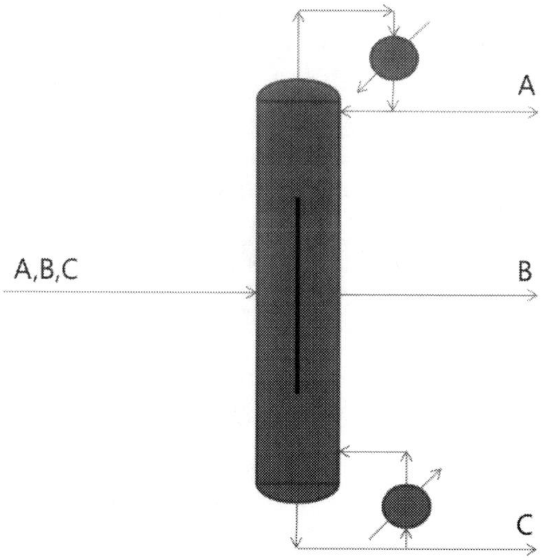

Process Intensification – Reaction Coupling

Another Process Intensification method is to couple separation and reaction to optimize the overall process.

An example would be using a reaction to convert a mixture with low relative volatility that would be difficult to separate with distillation into a mixture with that has high relative volatility that will separate much easier by distillation.

LPG Upgrading

The increase in North American shale oil and gas production has created a glut of NGL/LPG.

The initial wave of new ethane crackers will absorb some of the increased ethane production. However LPG (propane and butane) is primary used as fuel product . The increased supply is outstripping the demand leaving prices depressed.

The production of NGL in North America is expected to continue increasing.

LPG Upgrading

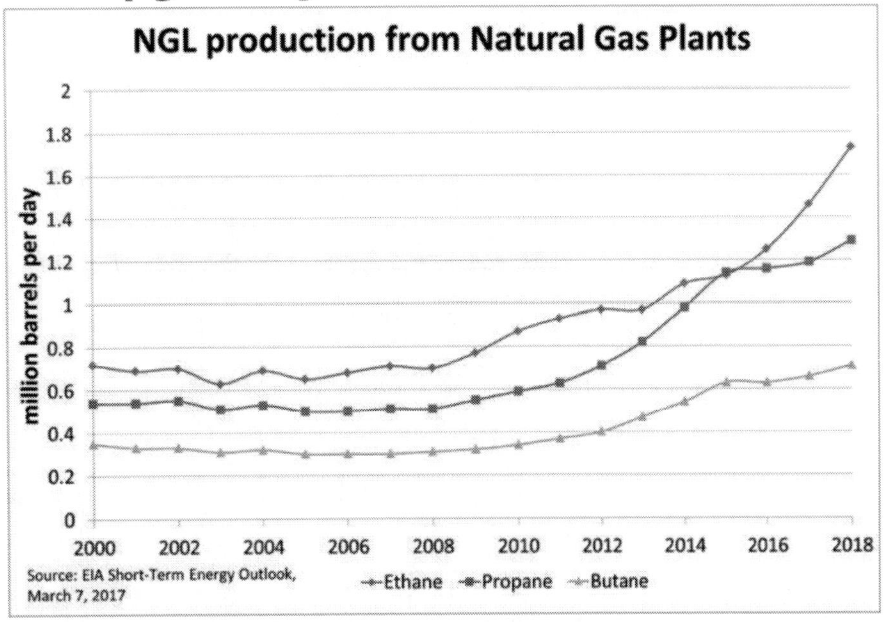

LPG Upgrading

The excess supply of LPG creates opportunities for use as a feed stock.

Propane dehydrogenation (PDH) and purification to polymer-grade propylene (PGP) is one process for upgrading propane that has received recent attention.

Polymer-grade propylene / polypropylene is a high value product. However, the process is capital and energy intensive.

Propane dehydrogenation (PDH)

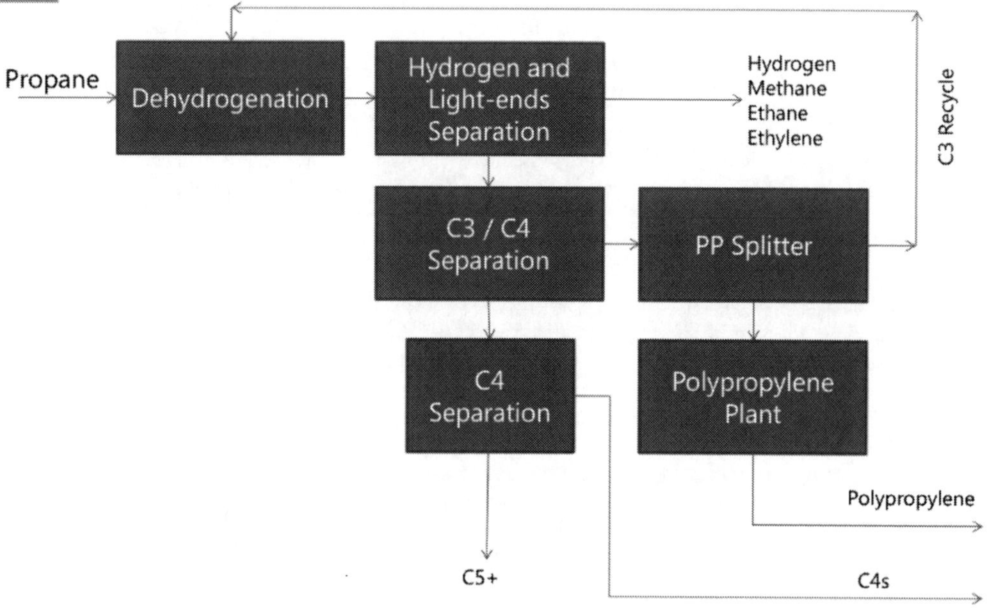

LPG Upgrading

Projects using this capital and energy intensive process are attractive when PGP and polypropylene prices are high enough . However, the overall market is limited and is unlikely to absorb the increases in propane and butane production.

Re-examination of existing technologies using a Process Intensification (PI) mindset leads to an alternate process for the creation of higher value products from propane and butane feeds stocks.

Propane and Butane to Liquids

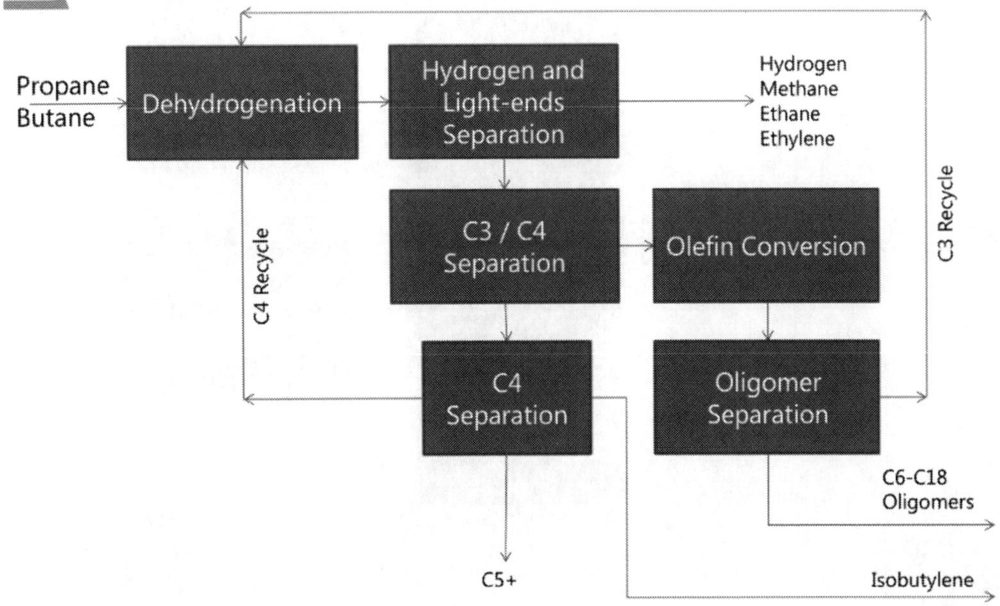

Propane and Butane to Liquids

In the alternate process olefin conversion is used to convert the C3 and C4 olefins into oligomers – molecules with a C6 to C18 carbon count or other lower volatility compounds.

This is the first PI step. The process for polypropylene production requires the distillation of the propylene to very high purity. This is very energy intensive because of the low relative volatility between propylene and propane.

Olefin conversion reactors can accept a mixed propane/propylene feed. The conversion of the propylene into lower volatility molecules makes the separation process much easier.

Propane and Butane to Liquids

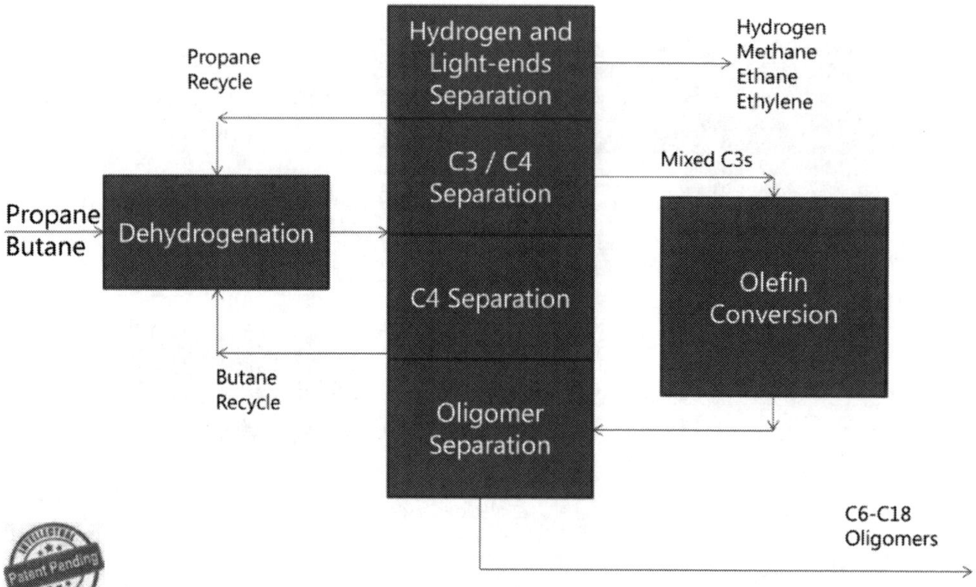

Propane and Butane to Liquids

The second PI step is to use DWC to integrate the distillation sequences and to configure the reactor to be an integrated side unit closely attached to the divided wall distillation column.

This configuration offers an attractive alternate for the utilization of excess propane and butane.

Patent Pending

Questions

????

More information will be presented in the Developments in C3-C5 Petrochemicals II

Converting LPG Into High-Value Products

Integrated process for dehydrogenation and conversion of propane and butane

Advisian

WorleyParsons Group

2017 AIChE® SPRING MEETING & 13TH GLOBAL CONGRESS ON PROCESS SAFETY

2017 AIChE® SPRING MEETING & 13TH GLOBAL CONGRESS ON PROCESS SAFETY

481018 Systematic Process Intensification Using Building Block Approach

Abstract:

Though research on process intensification has brought about many promising technologies including reactive distillation, divided wall columns, membrane reactors and task integrated columns, there is a lack of systematic methods for the identification and optimization of intensification alternatives at the flowsheet and equipment levels. Current superstructure-based process synthesis relies on pre-specified configurations and is unable to automatically construct novel equipment configurations for intensification. In this work, we depart from the classical unit operation-based representation of process units, flowsheets and superstructures and propose a new representation using fundamental building blocks which leads to a systematic method for simultaneous process design, synthesis and intensification. These building blocks can be associated with different process phenomena. An assembly of blocks of the same type obtains a classical unit, while an assembly of blocks with different types results in an intensified unit. This allows to systematically identify and incorporate many intensification pathways using a general block-based superstructure at the equipment and flowsheet levels. We obtain the intensified process by optimizing a performance metric for given raw materials and product specifications, material properties and bounds on flow rates. The overall intensification problem is formulated using a single mixed-integer nonlinear optimization (MINLP) model which is solved using commercial solvers. In this presentation, we will demonstrate the applicability of our approach using several process design and intensification case studies.

References:

[1] Demirel, S.E., Li, J., Hasan, M.M.F. Systematic Process Intensification using Building Blocks. Submitted to *Computers & Chemical Engineering*.

483476 Smart Manufacturing Business and Technology Practice

Abstract:

Smart Manufacturing (SM) uses the integration of next generation Operations Technology (OT) and Information Technology (IT) to realize significant untapped market opportunities. Opportunities accrue from a broader, more customized product space and accelerated dynamic, and precision manufacturing, drawing upon the cumulative impacts of every step across entire enterprise value and supply chains. Significantly increased energy, material and workforce productivity and improved environment sustainability are pathways to opportunities as well as outcomes of enterprise optimization. Smart Manufacturing (SM) is a term of 'practice' to just technology. It is in what way, at what speed, and how well data and next generation IT are applied that impact the future of individual manufacturers and their supply chains. Smart Manufacturing (SM) is therefore the opportunity-based and business oriented application of next generation advanced sensors, controls, platforms and modeling. Building meaning through a lens of practices that orchestrate capabilities to achieve business and operational objectives makes it possible to describe SM itself as set of practices that distinguish it from today's 40 year history of applying IT to manufacturing.

482647 Process Intensification for Reactions and Reactors

Abstract:

We will describe various concepts for intensifying catalytic reactions and reactors, including bifunctional materials and cascade reactions, fairly mature by now microtechnology, hybrids of reactors and separations, and multifunctional reactors including static micromixers and microheat exchangers (spatial domain), periodic operation (temporal domain), and use of alternate energy forms such as microwaves. These concepts will be elucidated with examples taken from biomass, shale gas, and ammonia chemistry.

483477 Process Intensification at Different Length-Scales

Abstract:

Process intensification (PI) concepts offer opportunities to enhance the energy efficiencies, and reduce the emissions and costs for the manufacture of chemicals, medicines and fuels. This talk will describe our use of PI concepts at different length-scales to design and demonstrate two process systems: a highly efficient, thermally integrated system for the production of hydrogen, and a nanoscale cascade catalytic system for the hydrogenation of CO_2 to value-added products. The hydrogen production system is designed to use latent from the vaporization of H_2O to control the axial temperature gradient of a reactor that couples autothermal reforming, water gas shift and preferential oxidation. Cascade systems use multiple catalytic species to facilitate multi-step reactions in a single vessel. A low temperature, heterogeneous cascade system is demonstrated for the selective hydrogenation of CO_2 to methanol and dimethyl ether via formic acid and methyl formate intermediates. Technoeconomic analyses indicate that these two modular systems could reduce the cost of distributed hydrogen production and CO_2 hydrogenation.

483079 Process Intensification in Multicomponent Distillation

Abstract:

Process intensification (PI) is an emerging concept which describes the design innovations that lead to significant shrink in size and dramatic boost in efficiency in a process plant. Distillation, which accounts for 90-95% of all separations in a chemical plant and ~3% of the world's energy consumption, is therefore a crucial component in PI. In the context of multicomponent distillation, we believe that PI stands for minimizing the number of pieces of equipment as well as total cost while maintaining high efficiency through innovative process synthesis. This will lead to compact, easy-to-operate, energy efficient and cost effective processes. Here, we consider several different levels of PI in multicomponent distillation, including: 1) Introducing thermal couplings that eliminates some or all reboilers and condensers to save energy and reduce capital cost; 2) Using thermal couplings strategically by identifying places where heat exchangers that use intermediate temperature utilities should be kept to enhance thermodynamic efficiency, as well as places where thermal couplings are preferred in order to provide substantial energy savings but incur no or minimal penalty in utility; 3) Performing simultaneous heat and mass integration among thermally coupled distillation columns to reduce the number of columns and heat duty requirement; 4) Conducting any thermally coupled distillation in only one single column shell using the form of a dividing wall column that is fully operable; 5) Being able to systematically derive numerous configurations that use 1 to (−2) column shells from a traditional (−1)-column configuration. Here, we demonstrate these aspects of PI through examples, illustrating the importance of PI in multicomponent distillation and how this concept can be utilized by process engineers to design cost effective and energy efficient distillation systems.

482646 Autothermal Process Intensification: Overcoming Heat Transfer Bottlenecks in Non-Equilibrium Chemical Processes

Abstract:

Heat transfer is the bottleneck in many chemical processes, limiting throughput for a reactor even when chemical reaction and mass transfer are very fast. Heat transfer can be enhanced through a number of strategies, often adding significantly to the capital cost of a project. However, as long as heat transfer is rate limiting, reactor capacity only scales as the square of reactor diameter compared to the cube of reactor diameter for volumetric chemical rate-limited reactors. This not only constrains throughput as reactors are built larger, but substantially eliminates the economic advantage of building them larger.

The heat transfer bottleneck can be overcome through *autothermal chemical processing*, which balances the energy demand of endothermic chemical reactions with energy supply from exothermic chemical reactions within an adiabatic chemical reactor. Although autothermal chemical reactors have been previously demonstrated for *equilibrium* chemical processes (prominent examples include methane steam reformers and biomass/coal gasifiers), we are developing it for inherently *non-equilibrium* processes. Identifying complementary endothermic and exothermic reactions that can occur simultaneously while preserving the same products of an externally heated (or cooled) chemical process is not trivial. However, the gains in *process intensification* would be substantial, increasing throughput and reducing capital and operating costs of a chemical process.

We have applied the principle of non-equilibrium autothermal chemical processing to the pyrolysis of biomass. Although the enthalpy for pyrolysis is relatively small (comparable to the enthalpy of evaporation for methanol) and operates at 500°C, throughput for a pyrolyzer is constrained by the rate that thermal energy can be transported into the reactor. This energy is conventionally supplied from hot flue gas conveyed through heat exchanger tubes or hot granular media like sand or steel shot mixed directly with the pyrolyzing biomass. Although these heat transfer systems add tremendous complexity to the pyrolyzers, they can sustain only modest thermal fluxes in practical pyrolysis systems, limiting the commercial prospects for pyrolysis energy systems.

Autothermal pyrolysis is achieved by admitting a small amount of oxygen to a fluidized bed reactor, which partially oxidizes biomass reactant or pyrolysis products to provide the enthalpy of pyrolysis. Experiments performed in a 15.4 cm dia. fluidized bed reactor capable of operating as either a conventional or autothermal pyrolyzer found no significant loss in bio-oil yield or quality for these two modes of operation at comparable biomass throughput. More importantly, by removing the heat transfer bottleneck of conventional pyrolysis, operation in autothermal mode allowed biomass throughput to be dramatically increased, reaching five times the capacity of the conventionally operated pyrolyzer. Our analysis indicates oxidation of non-condensable gas species released during pyrolysis provided about half the energy for pyrolysis while the remainder of the energy is thought to come from partial oxidation of lignin in the biomass and char in the pyrolysis products.

482648 Process Intensification and Modular Production for Converting Stranded Gas to Value-Added Products

Abstract:

In the United States and rest of the world, there are abundant gas resources which are either physically or economically stranded. These gases include flaring gas in shale oil field, refinery off-gas, coal-bed methane, shale gas in disadvantaged locations. Upgrading of these gases to value-added liquid products could reduce the demand on crude oil by about 20% in the United States.

On-site modular production of value-added products from stranded gases become attractive. Due to economy of scale, modular production typically doesn't compete well with conventional large scale production. As a result, process intensified modularity becomes important because it will enhance productivity, energy efficiency and flexibility that allows the modular production to be economically viable. What is process intensified modular production? It consists of novel apparatuses and techniques that, compared to those conventionally practiced today, are expected to bring dramatic improvements in manufacturing and processing, substantially decreasing the ratio of equipment size to production capacity. Most importantly, the process intensification should reduce energy consumption, waste production, and ultimately resulting in cheaper and sustainable technologies. Illustrated in this presentation are synergistic integration of the following unit operations: reactor design, heat management and separation. Process intensification sets the foundation for small modular production to achieve both energy efficiency economic production.

Indirect conversion of natural gas to liquid products (GTL) via syngas has been commercialized but it requires huge capital investment, varying from $15 to 20 billion for a 100-140 bbl/d plant. Direct conversion of natural gas without going through syngas route has been under investigation in the past 5 decades, but no commercial processes are practiced to date. A large number of studies have been published on the subject over the past 50 years.

This study emphasizes the direct conversion of short chain parafins (C_1-C_3) into aromatics and olefins using transition metal promoted ZSM-5 zeolite catalysts. Catalyst activity, selectivity, deactivation and regeneration of metal-promoted ZSM-5 zeolite catalysts will be discussed. We will introduce a new approach that employs low temperature plasma to intensify catalytic reaction for natural gas conversion. Under low reaction severity, this approach synergistically integrates plasma reaction chemistry with novel heterogeneous catalysis that decouples methane activation from catalytic surface reaction, shifting rate-determining step from methane activation (cracking C-H bond) to surface C-C formation.

In summary, this presentation illustrates experimental research on direct non-oxidative conversion of natural gas via process intensified production. The challenge in advance the fundamental science aspects presented in direct natural gas conversion is discussed. By comparing with commercial hydrocarbon conversion processes, the industrial perspective on direct methane conversion is illustrated. Specifically, the strategies on the selection of reactor configuration, heat management and catalyst regeneration are highlighted. These commercialization strategies along with technoeconomic analysis are critical for direct natural gas conversion to value-added chemicals.

483478 Reinventing Nuclear Chemical Engineering through Modular Process Intensification and Smart Manufacturing

Abstract:

The Department of Energy ("DOE") is responsible for the cleanup of the legacy of nuclear weapons production. Beginning with the Manhattan Project and extending over the course of the Cold War, DOE operated a massive industrial chemical complex to produce and separate nuclear materials. DOE utilized the best-in-class chemical companies to research, develop, design, build and operate the chemical complex. Liability law changes and changing business strategies caused these best-in-class chemical companies to exit the DOE businesses. Thus, chemical engineering essentially ceased to advance as a science and applied engineering field within the DOE complex. When the Cold War ended, the integrated operations of the complex were shut down and DOE began the costly long-term effort to clean up the waste resulting from the operations. Over $150B has been spent to date in pursuit of this mission, with the result that cleanup of 91 of the 107 major sites has been completed. With a baseline budget of $6B/year, the cost to complete the cleanup mission is on the order of $300-335B and that cleanup would extend for many decades after 2060. Both DOE and Savannah River National laboratory believe that technology innovation coupled with workforce development in the fields of modular process intensification and smart manufacturing have the potential to decrease capital, operating and final life cycle costs. This presentation will discuss some of the successes achieved to date and discuss SRNL's efforts to leverage other government investments to replace the loss of chemical engineering knowledge within the DOE Environmental Management and National Nuclear Security Missions.

483080 Process Integration and Intensification: Challenges and Opportunities in Renewable Bioproducts Including Pulp and Paper

Abstract:

The forest bioproducts industry today includes the pulp, paper, composite and allied chemical and other manufacturers. This segment of the US and North American manufacturing industry has been globally competitive for much of the last century because of the abundance of high quality forest resources as raw materials and highly optimized large scale processes such as the chemical recovery system in pulp mills. The focus of the industry upto now has been primarily on producing low cost cellulose fiber and fiber based products including building materials. This is a mature commodity industry with high capital costs and low profit margins. With increasing cost of raw materials and labor and operational efficiency and leveling the technology playing field for conventional fiber based products, the forest industry is increasingly looking to expand the revenue streams and profitability. As pulp and paper mills and forest products manufacturing units evolve into integrated forest biorefineries, they will produce a broader spectrum of chemicals and materials with higher value. The evolution towards diversification of products and materials while pursuing higher value and economic profitability presents tremendous challenges and opportunities for process integration and intensification. We expect process intensification to result in more compact, modularized processing and converting operations serving to improve profitability while better utilizing the raw material and energy inputs.

With significant expertise in the understanding, logistics, handling and processing of lignocellulosic biomass and sustainable forestry practices, the forest industry has much to offer in contributing to the future development of sustainable bioeconomy. The RAPID Renewable Bioproducts technical focus area is formed with the goal to develop, demonstrate and deploy innovative approaches and solutions of process intensification in the pulp and paper and other biobased industry improving energy efficiency, increasing productivity, improving environmental performance, minimizing waste and better utilization of waste streams, and the development of novel value added bio-products diversifying the product spectrum to enhance revenue. It is proposed to deploy a wide range of process intensification technologies based on membranes, reactive processes and advanced novel separations for pulping, bleaching, black liquor and chemical recovery, papermaking processes, reduction and better utilization of wastes, biorefininig including thermochemical, biological and biochemical processesand development of value-added bioproducts.

Specifically, the pulp and paper area will specifically aim to address the key goals of increased value through modularization and intensification of processes to increase energy efficiency, doubling energy productivity via reductions in capital costs and operating costs, enabling cost effective deployment of new technologies in existing manufacturing sites, and improved work force by focusing on better education of students. The project efforts will strive to address the strategic goals for the pulp and paper industry developed by Agenda 2020: to reduce the total input energy by 25%, increasing fiber yield by 5% on oven dry (O.D.) basis, reducing BOD/COD of effluents by 25% and reduction of the industry's overall capital intensity by at least 25%.

Kraft pulp mills and bleaching operations alone consume an estimated 562 trillion Btu. From the Agenda 2020 analysis, it is projected that breakthrough next generation pulping and bleaching technologies offer the potential for saving approximately $950 million per year in energy and raw material costs alone. Other specific metrics have been proposed and these include achieving energy savings in the pulping process of up to 10%, a target to increase the pulp yield by 4-6%, increasing the extent of delignification (targeting up to 15% increase in lignin removal) and obtaining an improved pulp strength and bleaching response. Eliminate scaling in chemical recovery and heat and mass transfer operations thereby improve the overall heat and mass transfer rates and energy consumption.

About 7% of pulp and paper energy usage, or nearly 164 trillion Btu, is used to remove water from black liquor in U.S. Kraft mills. It is estimated that integrating membrane-based black liquor concentration technology has the potential to reduce annual energy costs by $2–$3 million typical for a typical 2,000 TPD mill.

It is estimated that paper drying alone consumes about 20% of the total energy consumed by pulp and paper industry. It is targeted to increase the moisture content going into the drying stage to approximately 60-65% solids from the current 40-45% solids; increase the drying rate by about 10%; decrease the energy consumption in the pressing and drying section by about 10%. The above target, if successful, could save 80 trillion Btu per year in the United States, or $240–$400 million.

Process water usage in pulp and paper manufacturing is another key area. Here the target goal is to reduce the water usage by 50% by focusing on reducing water use and reuse during papermaking on the paper machine and reusing effluent water back in the manufacturing process.

Another key component as part of an integrated biorefinery is aimed at improved utilization of biomass feedstock and waste streams from pulp and paper manufacturing including the development of cellulose nanomaterials and value-added bioproducts. This is aimed at addressing the strategic goal of producing additional revenue and improving the overall economic viability and sustainability of the pulp and paper industry.

Working closely with the pulp and paper industry it is important to integrate the learning from the research and demonstration projects in undergraduate and graduate curriculum specifically in courses in the areas of pulp and paper engineering, process engineering, biomass chemistry and conversion, reaction engineering, separation process principles, unit operations laboratory, and engineering design. In addition it is critical to incorporate appropriate technology developments in outreach and continuing education efforts including certificate programs geared towards industry professionals and workforce training and development.

RAPID working together with Agenda 2020 and the pulp and paper and other biobased industry, academia and national labs is well positioned to address these challenges using a strategic integrated approach involving key technical and application focus areas. Pulp and paper area is fortunate to have established academic institutions and national labs bringing significant expertise and research and development capabilities including pulping, bleaching and paper making unit operations, biomass characterization and conversion equipment, wide spectrum of analytical characterization and instrumentation including physical testing for pulp and paper, pilot scale pulping and bleaching equipment, black liquor characterization and separation, paper machines and biorefining processing capabilities and waste water treatment capabilities. In addition through our industrial partners' larger scale pilot and commercial scale processing equipment can also be used for large scale demonstration and commercialization.

483803 "An Overview of Process Intensification Research at Newcastle University (UK)"

Abstract:

A brief overview of research within the Process Intensification Group at Newcastle University (UK) will be given. This will encompass some generalities about process intensification, and brief descriptions of key novel process technologies and their applications at Newcastle. Currently the main focii of research are: oscillatory baffled reactors, rotating packed beds, spinning disc reactors, non-thermal plasma processing and heat pipe technology.

483581 Opportunities and Challenges for Process Innovation and Intensification: An Industrial Perspective

Abstract:

Despite more than a century of developments in chemical processing, large opportunities for energy and atom efficiency remain. Increasing population means that demand for products of the industry will continue to grow, posing a challenge to shrink the environmental footprint per unit of production.

This talk will examine some of the process improvement opportunities of interest to Dow and how we have attempted to address them. In addition to highlighting some success stories, the talk will present some of the barriers we encounter when attempting to innovate and intensify process technologies.

483582 Intensification of Processes yielding Renewable Bio-Products: Theory - Experiments - Field Applications

Abstract:

The historical journey of Process Intensification is first reviewed, and placed in context of prior research activities. Then the potential role of Process Intensification and the AICHE RAPID Institute in the development of novel processes for the creation of Renewable Bio-Products will be outlined. Emphasis will be given to a wide array of theory tools that can be brought to bear for the synthesis of novel processes yielding renewable bio-products, experiments that can be used for the validation of these theory tools, and the practical development of proposed new technologies, and a number of field applications that can serve as demonstration vehicles for the proposed technologies.

AUTHOR INDEX